SAMPLE SAKURA VILLA

格居·樱花墅

托尼 编著

辽宁科学技术出版社
·沈阳·

图书在版编目（CIP）数据

格居·樱花墅 / 托尼编著 .—沈阳：辽宁科学技术出版社，2012.9
ISBN 978-7-5381-7550-9

Ⅰ．①格… Ⅱ．①托… Ⅲ．①别墅—室内装饰设计—图集 Ⅳ．① TU241.1-64

中国版本图书馆CIP数据核字（2012）第138769号

出版发行：辽宁科学技术出版社
（地址：沈阳市和平区十一纬路29号 邮编：110003）
印 刷 者：深圳大公印刷有限公司
经 销 者：各地新华书店
幅面尺寸：230mm×305mm
印　　张：49
插　　页：4
字　　数：400千字
印　　数：1～2000
出版时间：2012年9月第1版
印刷时间：2012年9月第1次印刷
责任编辑：郭媛媛
封面设计：ANYA CAI
版式设计：ANYA CAI
责任校对：徐　跃

书　　号：ISBN 978-7-5381-7550-9
定　　价：398.00元

联系电话：024-23284356　18604056776
邮购热线：024-23284502
E-mail:purple6688@126.com
http://www.lnkj.com.cn

PROFESSIONAL FOCUSED FASHIONABLE

专业 专注 时尚

THE DESIGNS CONTAINING THE SPIRIT "RETURN TO LIFE" WOULD SHOW PEOPLE THE PUREST AND BEST DESIGN LANGUAGE.

这些设计所包含的"回归生活"的精神,向人们展示出最纯粹、最美好的设计语言。

PREFACE 前言

In virtue of deduction by different designers, the same indoor space will come to various patterns. The designers learn well about expression of space value and desire from space, and make the cold space full of sweat and warm life atmosphere nothing less than an exploration for life art.

Different inner designs reflect the different culture and style in different regions, and reveal connotation and attainment of owner. As an excellent inner designer, he will not only plays a role in forwarding and expressing the updated design conception, but explores the future to find all possibilities of household life for satisfying need from individuality.

一样的室内空间,经过不同设计师的演绎,便能产生完全不一样的格局。设计师充分理解空间价值的表达和空间自身的渴求,让原本冰冷的空间充满温馨与温暖的生活气息,这不亚于一次对于生活艺术的探索。

不同的室内设计不仅能反映出不同地域的文化与风格,还能呈现居者的内涵素养。作为优秀的室内设计师,不只起到传达和表达最前沿设计理念的作用,还要向更远的未来探索,寻找家居生活的各种可能性,以符合个性化的需求。

2012

Organizer Century Sakura(HK) Entertainment & Media Co.,Ltd
编写单位 香港世纪樱花娱乐传媒有限公司

Publisher 出版人兼总监　Tony Lin
Advisor 顾问　Andy Du
Art Director 美术指导　Anya Cai
Editor-in-Chief 编著　Tony Lin
Executive Editor 执行编辑　Gavin Gong & Christina Dong
Senior Graphic Designer 资深美术编辑　Anya Cai
Graphic Designers 专题美术编辑　Elaine Li
Market development 市场拓展　Fernando Wang

Hong Kong Office 香港
ADD: 香港九龙尖沙明广东道5号海港城海洋中心8楼827室
Unit 827,8F,Ocean Center,Harbour city,5Canton Road,Tst Kowloon,Hong Kong
TEL: +852 28763793
FAX: +852 21296308

Beijing Office 北京
ADD: 北京市朝阳区建外大街四号建外SOHO 2号楼2003室
Room 2003,SOHO 2 Building,4Jianguomenwai Street,Chaoyang District,Beijing
Zip Code:100022
TEL: +86 10 51266666
FAX: +86 10 51288866

Shenzhen Office 深圳
ADD: 深圳市福田区竹子林求是大厦东座2903室
2903,east wing,Qiushi Center,Zhuzilin, Futian District,Shenzhen,Guangdong
Zip Code:518000
TEL: +86 755 82822480
FAX: +86 755 82822481

Chengdu Office 成都
ADD: 成都市锦江区香格里拉写字楼8楼803室
Room 803,Shangri-La Office Building,Jinjiang District,Chengdu
Zip Code:610000
TEL: +86 28 84438072
FAX: +86 28 84438073

Wenzhou Office 温州
ADD: 温州乐清市宁康西路388号
388 Ningkang Road W., Yueqing,Wenzhou
Zip Code:325000
TEL: +86 577 61576666
FAX: +86 577 61510066

Wuhan Office 武汉
ADD:武汉市武昌区临江大道96号武汉万达中心写字楼
Room 2003,Wanda centet,96Linjiang Road,Wuchang District,Wuhan
Zip Code 430000
TEL:+86 27 86728088
FAX:+86 27 86728288

www.centurysakura.com

Statement
Copyright(C) HK Century Sakura. All rights reserved. Without our prior written consent, no graphic or text contained in this publication may not be reproduced, reprinted, distributed, or otherwise used in any form for any purpose.

声明
版权归世纪樱花（HK Century Sakura）所有。未经本公司书面许可，不得以任何目的、以任何形式或手段复制、翻印、传播，或以其他任何方式使用本书的任何图文。

About Sakura Villa
樱花墅简介

Sakura Villa is specialized in villa interior design for frontline brands by introducing fashion elements into space design to interpret the ideal of our life. Our business involves interior and exterior design and operation management in the aspects of architecture planning, office space, commercial space and parking space. Our projects once involve design and decoration of shopping center, hotel, restaurant, coffee house, premise marketing center, sample apartment, chamber, SPA, designer store and office building.

Sakura Villa focuses on furnishings, materials that best go along with the environment, leading fashion taste and matching customer requirement to bring a unique feeling and touchable comfort in every space. To keep in line with international design ideas, we invited several well-known European and American designers and sent our designers overseas to study once at a time to strengthen international communication and synchronize Sakura Villa design idea and international advanced strength, introducing the newest interior design ideas to further meet requirements of people who pursue higher life quality.

With our own strength and foresight, international top-notch elite team of Sakura Villa walked in the front of design industry by their superior design ideas and acute design sensitivity, and created personal furnishing service system integrating design, operation, furnishing customization and post accessory. We comprehend and value commercial value of design; we pursue the combination of function, technology and art; we emphasize sense of worth and value-added effect brought by products to our customers.

樱花墅专业于豪宅家装设计，专注于服务一线品牌，将时尚导入空间设计，诠释品质生活的理念。服务领域涉及展示空间、办公空间、商业空间、居住空间的室内外装饰设计及工程施工管理的各个方面。过往项目包括购物中心、酒店、餐厅、营销中心、样板间、会所、SPA、品牌店及写字楼的设计装饰。

樱花墅关注最时尚的家具、材质，引领时尚品位，精准定位客户的心理需求，为每一个空间营造独特的感觉和可以触摸的舒适。为了时刻保持设计概念同国际接轨，我们直接邀请了数名欧美著名设计师参与设计，并分期分批派送设计师出国考察，以此加强国际化的交流，使樱花墅的设计理念与国际先进水平同步，将世界最新的室内环境设计理念融进我们的设计方案，以满足追求更高生活品质的人们的需求。

凭借自身实力和超前意识，樱花墅国际顶尖设计精英团队以超前的设计概念、灵敏的设计嗅觉走在设计领域的最前端，创立了集设计、施工、家具定制、后期配饰于一体的家居全程私人服务体系，理解并重视设计的商业价值，追求作品在功能、技术和艺术上的完美结合，注重作品带给客户的价值感和增值效应。

 P016
 P032
 P052
 P068

 P150
 P172
 P190
 P202

 P286
 P294
 P302
 P314

 P074
 P100
 P144
 P136

  P232  P246  P258  P272

 P324  P334  P346  P354

CONTENTS
目录

Forest of Cubes Xiamen Center Kingdom Sales Office	立方体密林 厦门城立方售楼处	P016
Element Shows Low-key Luxury Type No. 505, Xiamen Center Kingdom	元素呈现低调奢华 厦门城立方 505 号户型	P032
The Philosophy of Details Type No.502, Xiamen Center Kingdom	倾情演绎细节哲学 厦门城立方 502 号户型	P052
Beauty of Magnificence Lobby of Building 36, Center Stage	恢弘气度之美 金域中央 36 栋大堂	P068
Luxury as Dubai L24/F, Building 36, Center Stage	迪拜式奢华 金域中央 36 栋 24 层楼王	P074
Life in Classical Europe Kunming Vanke – Baisha Runyuan Villa	古典欧式的新生命力 昆明万科·白沙润园别墅	P100
Return to Chinese Classical Zhongliang Luofu Mountain & Water Life Ecology Town, Huizhou	回归中式古典 惠州罗浮山中量山水人间生态城	P114
A Tasteful Life A2, Building 2, Oriental Vision Phase I	有品位的生活 锦绣御园一期 2 号楼 A2	P136
Get Immersed in the Age A1, Building 3, Oriental Vision Phase I	沉醉于岁月的沉淀 锦绣御园一期 3 号楼 A1	P150
Art of Layered Space Xiamen Yongjian – Dingshang Capital Villa	层次空间艺术 厦门永建·顶尚商墅	P172
Charm Sources from Culture Unit 2-A, Building 2, Shum Yip Yuyuan	文化感传承风韵 深业御园 2 号楼单位 2-A	P190
Fun about Balance Kunming Vanke – Baisha Runyuan Type C	有关平衡的乐趣 昆明万科·白沙润园 C 户型	P202
Tell an Expected Story Changcheng – Longwan	叙说一个向往的故事 长城·珑湾	P212
Simple Black & White A1, Building 1, Oriental Vision Phase I	简约黑白协奏 锦绣御园一期 1 号楼 A1	P232

Simple Elegance of Blue Unit 2-B, Building 2, Shum Yip Yuyuan	蓝色简约之雅 深业御园 2 号楼单位 2-B	P246
From Dream to Reality Shangdongwan Type B-A	梦想照进现实 上东湾 B-A 户型	P258
A Space with Sanskrit Kunming Vanke – Baisha Runyuan Type D	梵音弥散的空间 昆明万科·白沙润园 D 户型	P272
The Art of Line Kunming Vanke – Sales Office of Baisha Runyuan	线形的艺术 昆明万科·白沙润园售楼处	P286
The Magic of Mosaic 1-C, Unit B, Building 1, Shum Yip Yuyuan	马赛克幻觉魔法 深业御园 1 号楼 B 单位 1-C	P294
The Blue of Ocean Kunming Vanke – Baisha Runyuan Type E1	海的蓝在此飞扬 昆明万科·白沙润园 E1	P302
Dramatic Beauty with High Contract 1-B, Unit B, Building 1, Shum Yip Yuyuan	强对比的戏剧性美感 深业御园 1 号楼 B 单位 1-B	P314
Make Everything at Ease B2, Building 1, Oriental Vision Phase I	让一切回归安然 锦绣御园一期 1 号楼 B2	P324
Life in False or True 1-D, Unit B, Building 1, Shum Yip Yuyuan	虚实之间理解生活 深业御园 1 号楼 B 单位 1-D	P334
Fun of Tenderness Shangdongwan Type B-F	柔和的趣味 上东湾 B-F 户型	P346
Concept Space	概念空间	P354

SAKURA VILLA
樱花墅

SUBLIMATION
纯净
EMPHASIZING
特写
SYMBO
象征

FOREST OF CUBES
立方体密林

Xiamen Center Kingdom Sales Office
厦门城立方售楼处

1 SAKURA VILLA
樱花墅

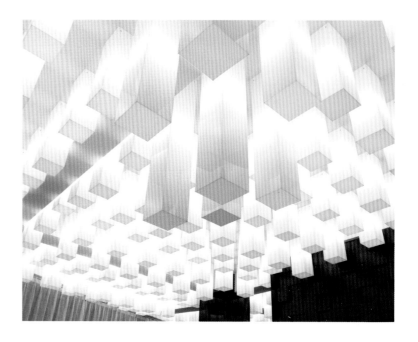

FOREST OF CUBES
立方体密林

Project name Xiamen Center Kingdom Sales Office	项目名称 厦门城立方售楼处
Design theme Modern style	设计主题 现代
Building area 301 sqm	建筑面积 301 ㎡
Flat Type Lobby, Negotiation Area, VIP Area, Financial Room, Sales Office, Resting Room	户型 大堂、洽谈区、VIP 区、财务室、销售办公室、休息室
Developer Xiamen Chongjun Real Estate Co., Ltd.	开发商 厦门崇峻置业有限公司

Here you may not fall in with the crowd, or make any excessive demands; here a new design concept is adopted, without luxurious wallpaper, but only white lamp boxes and brown wooden surfaces. The well arranged lamp boxes and wooden surfaces form the base of the space, and elements are constantly changed with same materials in different spaces, emphasizing an overall feeling. Every corner of the space is the essence of the thoughts, and the purity of arts. The "humorous toys" are the first step of the sublimation of space, and the smooth and simple lines give the space more stories. The marble reception desk with irregular shape is the symbol of nobleness, and the humps on the ceiling and wall, let along their function, is the presentation of art. The curtain is well allocated with the entire space, and when swinging, it seems like the breeze from a far away country touching the Barcelona tables and chairs.

这里不随波逐流，不刻意追求。这里引领设计新概念，没有华丽的墙纸，只有白色的灯箱、红褐色的木饰面。排列错落的灯箱以及木饰面形成基调不同的空间，相同的材质、不断转换的元素，强调整体感，每一处都是思维撞击后集结的精华，是纯粹的艺术。"幽默小人"在空间内的存在是艺术升华的第一步，流畅的线形，既简单又灵动，凭空增添了太多的故事性。异形的切割大理石接待桌，是沉淀与高贵的体验，天花和墙面的突起，具有足够的艺术说服力，更不必说功能了。窗帘与整体风格的搭配，在摇摆中，仿佛接纳来自遥远国度的风，拂过散漫错落的巴塞罗那桌椅。

THE PURITY OF ARTS
纯粹的艺术

UNDISTURBED ENVIRONMENT

静谧的环境

It is a living room to have a free talk with guests, as well as a rest room. That is in virtue of the characteristics of undisturbed environment.

这是与客人畅谈的客厅，也是可以休憩的休息室，全因这份安静的环境特质。

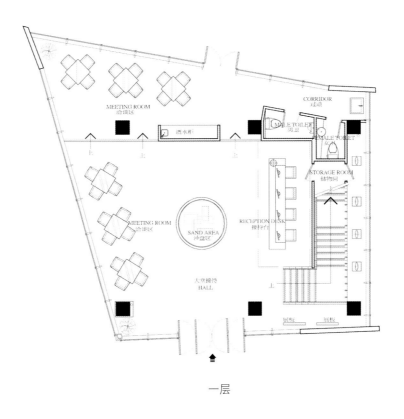

一层

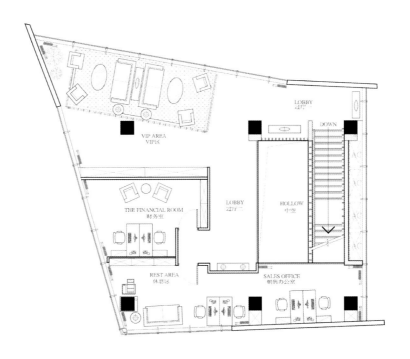

二层

SAKURA VILLA
樱花墅

CURVES
线条

GLOSSINESS
光泽

ELEGANCE
优雅

ELEMENT SHOWS LOW-KEY LUXURY
元素呈现低调奢华

Type No. 505, Xiamen Center Kingdom
厦门城立方 505 号户型

2 SAKURA VILLA
樱花墅

ELEMENT SHOWS LOW-KEY LUXURY
元素呈现低调奢华

Project name Type No. 505, Xiamen Center Kingdom	项目名称 厦门城立方505号户型
Design theme Chanel	设计主题 香奈儿
Building area 158.7 sqm	建筑面积 158.7㎡
Flat Type 4 bedrooms, 2 living rooms, 1 kitchen and 2 restrooms	户型 四房二厅一厨二卫
Developer Xiamen Chongjun Real Estate Co., Ltd.	开发商 厦门崇峻置业有限公司

Curves usually show tenderness and elegance. In this design, the designer uses a great number of curves to show a dynamic elegance. By making using of the hard materials with high glossiness and the different between hardness and softness, the entire space becomes colorful and changeful. The reason why this design is successful is because the curves are well used to enhance the expression of the space, representing the theme of the design. The black and white check pattern in the wall of the living room perfectly presents the soul of black and white of Chanel; the black and white lines in the wall of the restroom in living room, the white bathtub and the transparent articles in it are harmonious with each other, both classic and modern; in the kitchen, the green and white are well collocated with each other, with the customized overall cabinet in it, the noble feeling is naturally represented; the bedrooms are romantically decorated with silk like texture, achieving the nobleness and elegance as a swan.

曲线能显示出柔和与优雅。该案中，设计师大量运用曲线，流露出灵动与优雅；使用光泽度较高的坚硬感材质，通过刚柔之间的对比和微差，使整个空间构图富于变化。这个设计的成功之处就是运用曲线来加强空间的表现力，并以此表达设计主题。双C、黑白方块错落的客厅墙面装饰图案，完美演绎香奈儿黑与白的精髓；客卫黑白条纹相间的墙面、纯白的浴缸、剔透的器具极致融合，既古典又现代；厨房里，绿意与白色相辉映，量身定制的整体橱柜设计，尊崇品质不言自喻；主卧、客卧空间明亮而富有情调，如丝绸般的品质，成就白天鹅般的高雅尊贵。

THE SOUL OF CHANEL

香奈儿的灵魂

The black and white check pattern in the wall of the living room perfectly presents the soul of black and white of Chanel.

黑白方块错落的客厅墙面装饰图案，完美演绎香奈儿黑与白的精髓。

ELEGANT AND DISTINGUISHED
高雅尊贵

NOBLENESS AND ELEGANCE

高贵优雅

The bedrooms are romantically decorated with silk like texture, achieving the nobleness and elegance as a swan.

主卧、客卧空间明亮而富有情调，如丝绸般的品质，成就如白天鹅般高雅尊贵。

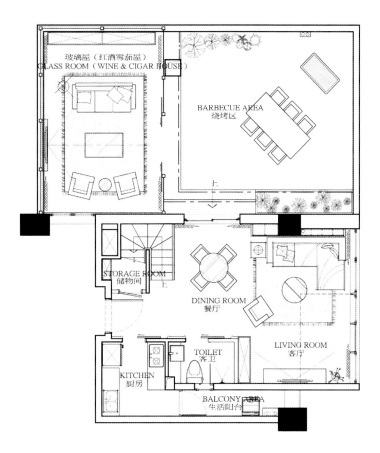

一层

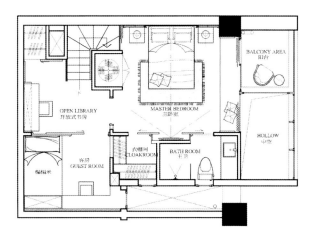

二层

SAKURA VILLA
樱花墅

ABSTRACT
抽象

STEADY
坚定

DEMEANOUR
风度

THE PHILOSOPHY OF DETAILS
倾情演绎细节哲学

Type No. 502, Xiamen Center Kingdom
厦门城立方 502 号户型

THE PHILOSOPHY OF DETAILS
倾情演绎细节哲学

Project name Type No. 502, Xiamen Center Kingdom	项目名称 厦门城立方 502 号户型
Design theme Hermes	设计主题 爱马仕
Building area 90.7 sqm	建筑面积 90.7 ㎡
Flat Type 3 bedrooms, 2 living rooms, 1 kitchen and 3 restrooms	户型 三房二厅一厨三卫
Developer Xiamen Chongjun Real Estate Co., Ltd.	开发商 厦门崇峻置业有限公司

3
SAKURA VILLA
樱花墅

If an abstract feeling can be materialized, it must come from the scale in large space or the thought in every detail. The generous scale and unique detail can be found everywhere in the space. The matador carpet, dark brown glasses, and colorful Hermes scarves in between have brought elegance to the space while making it more stable and steady. To compare with the simple but generous brown color adopted by the entire space, the colors of the decorations are more vivid, including white, black and red, creating the rich color with strong conflicts with the brown color, so that the layers of the space become richer. In the light and shadow, the warm feeling of the living room is created. The handmade delicate porcelains, which beautiful shape and high class material, create the elegant dinning environment; in the study room, different collections and books are placed in the yellow brown bookshelf, full of cultural feelings under the lamp; in the bedrooms, all articles are seem to be artworks with romantic arrangement and well selected material, showing the demeanour of a noble family.

如果抽象的感觉可以被物化，那么一定是载于大空间的尺度和小细节的心思。整个空间，大气的尺度和独特的细节处处可见，斗牛士的地毯，茶色的银镜，穿插色彩鲜艳的爱马仕丝巾，为空间增添雅致的同时，多了一份成熟稳重。相对于整个空间所采用的简约大气的咖啡色系，饰品颜色更为跳跃，白、棕、黑、红混合的醇厚色彩，与咖啡色形成对比，空间层次更为丰富。在绰约光影里，营造出客厅温暖的美学格调。手工制作的精密而细致的瓷器，造型优美，材质高档，铸就优雅用餐氛围；书房里，收藏品、各种书籍错落于棕黄色的书架间，在灯光的点缀下，文艺味道十足；主卧、客卧里，情调布置，精良选材，每件物品都是艺术品，尽显名仕风范。

WONDERFUL TECHNIQUES

完美的技术

To compare with the simple but generous brown color adopted by the entire space, the colors of the decorations are more vivid, including white, black and red, creating the rich color with strong conflicts with the brown color, so that the layers of the space become richer.

相对于整个空间所采用的简约大气的咖啡色系，饰品颜色更为跳跃，白、棕、黑、红混合的醇厚色彩，与咖啡色形成对比，空间层次更为丰富。

RICH COLOR

The generous scale and unique detail can be found everywhere in the space.

整个空间,大气的尺度和独特的细节处处可见。

五彩缤纷

NOBLE FAMILY
贵族家庭

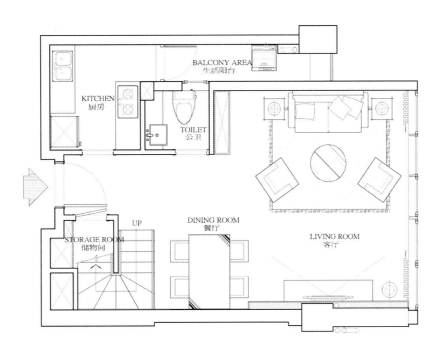

一层

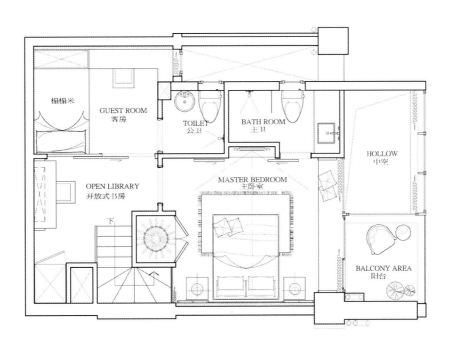

二层

SAKURA VILLA
樱花墅

NEOCLASSIC
古典

DELICATE
优美

POWERFULLY
魄力

BEAUTY OF MAGNIFICENCE
恢弘气度之美

Lobby of Building 36, Center Stage
金域中央 36 栋大堂

4
SAKURA VILLA
樱花墅

BEAUTY OF MAGNIFICENCE
恢弘气度之美

Project name Lobby of Building 36, Center Stage	项目名称 金域中央 36 栋大堂
Design theme Modern Luxury	设计主题 现代奢华
Building area 615 sqm	建筑面积 615 ㎡
Flat Type Lobby, Sub-lobby, Chess Room, Gym, Resting Area	户型 大堂、副大堂、棋牌区、健身区、休息区
AreaDeveloper Dongguan Kingon Real Estate Co., Ltd.	开发商 东莞市金众房地产有限公司

The large lobby seems rich and magnificent with the large crystal ceiling lamp in it. Under the luxurious lamp light, it seems like a curtain of time and space has been opened. The light of the crystal ceiling lamp can be reflected by both the marble floor and the silver folding screen, emphasizing the atmosphere of luxury. With the ceiling lamp as the focus of vision, the reflection of the lamp itself sets of the brilliant crystal, achieving an effect that only by making use of the characteristic of the material will the sense of layer be created. This is a discussion of space relation, in which the designer tries to create an atmosphere with an extreme imaging making method, presenting the significant vigor to the visitors when they step into the door. The linkage with other functional spaces makes the space meaningful, well representing a perfect combination between the comfortableness and fine arts, and that under the luxurious look, the practical nature is presented.

庞大的大堂因为大盏水晶吊顶的存在而显得异常饱满，华贵的灯光落下，仿若拉开了另一重时空的帷幕，运用大理石本身的光泽，水晶吊灯的折射光，水银镜屏风的反射感，突出豪华气氛。以中心的吊灯作为视觉的焦点，吊顶本身的反射效果衬托出夺目的水晶，仅是运用材料的材质特征就创造出不同的层次感。这是关于一种空间关系的探讨，设计师试图在利用一种极致的构图手法来营造氛围，将恢弘的气度在入门之时向来者呈现。而其他功能空间的一线式连接，使得空间格局充满了可以探索的意味，深刻体现了现代功能舒适性与审美艺术的高度结合，在奢华的外表之下不离务实的本质。

SAKURA VILLA
樱花墅

LUXURIOUS
奢华

NOBLE
高贵

REFINED
精致

LUXURY AS DUBAI
迪拜式奢华

L24/F, Building 36, Center Stage
金域中央 36 栋 24 层楼王

5 SAKURA VILLA
樱花墅

LUXURY AS DUBAI
迪拜式奢华

Project name	项目名称
L24/F, Building 36, Center Stage	金域中央 36 栋 24 层楼王
Design theme	设计主题
Neoclassic	新古典
Building area	建筑面积
538 sqm	538 ㎡
Flat Type	户型
5 bedrooms, 3 living rooms, 1 kitchen and 4 restrooms	五房三厅一厨四卫
Developer	开发商
Dongguan Kingon Real Estate Co., Ltd.	东莞市金众房地产有限公司

In this design, the sense of correspondence of space axis has been enhanced. According to the characteristics of the space, based on the choice of exhibits, and featured with the luxurious image, the key of design of the space has been formed. The step-in space is guided by the axis, which unifies the entire space and at the same time forms the different layer of the space. The axis, namely the order and layer relation of the space are well organized. At the junctions of certain axis, the golden color is adopted to further clear the layers of the space as well as the relations between them, and at the same time, more powerfully show the meaning of the word "luxury". The entire space adopts the neoclassic elements to show the features of luxury and elegance. The decorations indoor adopt ancient structures and color to emphasize the magnificent atmosphere of the space. The designer is proved to be good at making use of the luxury in the choice of the materials. The golden champagne color and the refection of mirror surface are adopted in the design to present the luxury of the space.

本案在设计中更强化了空间轴的对应意识，针对空间的特征、陈设品的选择，以华丽的形象语言为特征构成空间的设计要点，层层推进的空间，以轴线为引导，将整个空间区域规整为一体，又形成各个区间的层次对比。轴线关系，即空间的序列关系和层次组织。在轴线关系的某些空间区域交汇处，金色的介入，又使得整体空间的层次与虚实关系，对奢华一词的演绎，更为清晰有力。整个空间运用新古典的元素来展现豪华与雅致的特色，室内空间的装饰品采用古典的构建和色彩突出宏大的空间气氛。在材质的运用上体现了设计师对华贵气质的把握能力，运用香槟金色与镜面的反射特质来体现空间的豪华。

LUXURIOUS IMAGE

奢华映像

LUXURY AND ELEGANCE

奢华高雅

In the living room, the graceful and fine stone ground with wooden grain in Italian style, silk carpet in splendid colors. The soft colors, smooth touch and wonderful techniques reveal the nobility of great house.

客厅优雅细腻的意大利木纹石地面，色彩斑斓的真丝地毯，色彩柔和、触感轻柔、巧夺天工的技艺水平，尽显大宅风范。

ARDENCY AND TOUGHNESS

热烈张扬

The application of noble spotted grain adds more ardency and toughness. The culture wall with strong art atmosphere secretly improves the culture sense, European style ceiling lamp makes the whole space modest and elegant.

贵气十足的斑纹的运用增添了热烈张扬的气质,艺术气息浓厚的文化墙悄悄提升了客厅的文艺氛围,欧式吊灯的运用则使整个空间内敛而典雅。

JUMPING MUSICAL
跳跃音符

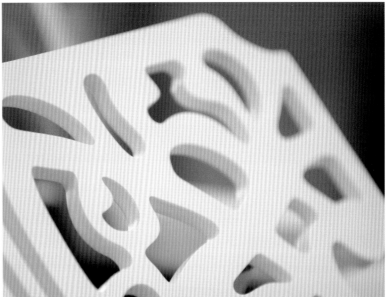

The ease piano corner endows luxury and lovely showing noble position and quality of the owner.

娴雅的钢琴角落,释放出极具质感的华美与感性,彰显了主人的尊贵身份和品质追求。

Wine and cigar provides with quiet study a suitable room for each side of life.
泛着岁月沉淀的红酒雪茄吧，静谧的书房，为人生的每一个侧面都提供了场合空间。

SUITABLE SPACE

相宜空间

金域中央 36 栋大堂

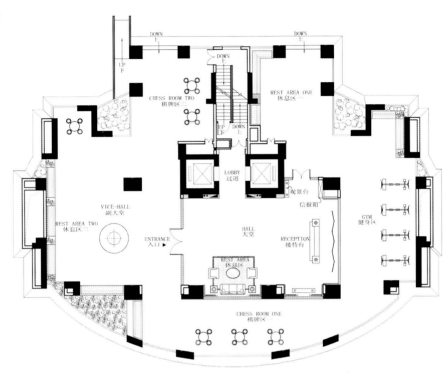

金域中央 36 栋 24 层楼王

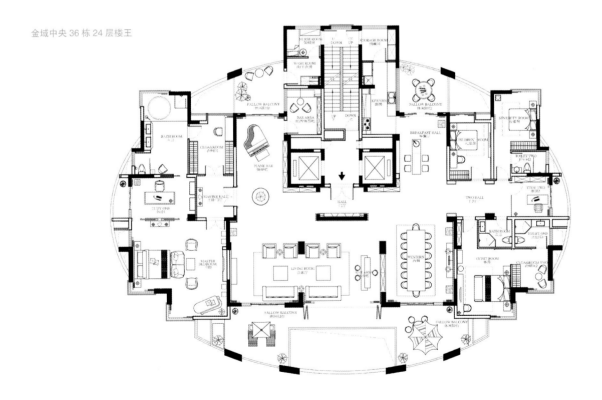

SAKURA VILLA
樱花墅

NEOCLASSIC
新古典

SYMMETRY
融汇

CLASSICAL
传统古典

LIFE IN CLASSICAL EUROPE
古典欧式的新生命力

Kunming Vanke–Baisha Runyuan Villa
昆明万科·白沙润园别墅

LIFE IN CLASSICAL EUROPE
古典欧式的新生命力

6 SAKURA VILLA
樱花墅

Project name Kunming Vanke – Baisha Runyuan Villa	项目名称 昆明万科·白沙润园别墅
Design theme Neoclassic	设计主题 新古典
Building area 614 sqm	建筑面积 614 ㎡
Flat Type 8 bedrooms, 4 living rooms, 1 kitchen, 6 restrooms and 2 garages	户型 八房四厅一厨六卫二车库
Developer Kunming Vanke Real Estate Co., Ltd.	开发商 昆明万科房地产开发有限公司

Based on neoclassic and adopting some ancient languages, the sense of separation created by luxurious space can be reduced and at the same time, by making use of the symmetry method in neoclassicism, the concept of the neoclassic design has been re-explained in an ancient manner, giving people the feeling of being stable and steady. Depending on the displays that are lower than the eye sight, the sense of layer of the space can be enhanced, and at the same time the completion and extension of the space can be maintained. The decoration is not overused to avoid negative influence to the magnificence of the space. The low-key luxury and extending smoothness are the theme of the design, based on which the designer creates a luxurious living space and maintaining ancient and fashion at the same time. In this design, the complication in ancient style is given up while the luxury is adopted, decorated with European style elements: whether the furniture, curtain, carpet or other decorations with the same style, or the traditional dinnerware, tea set or other articles full of value of collection and appreciation, have brought the noble and elegant breath to the living environment, and represented the vitality and power of classical European style in China.

以新古典为基调，融合部分复古语言，降低奢华空间造成的疏离感，同时利用古典主义对称的手法，以怀旧的方式重新诠释新古典设计的概念，给人沉稳自在的感觉。利用低于视平线的陈设将空间区隔，增加空间的层次感，同时保有大尺度空间的完整性及延伸的效果，不过度刻意地使用装饰符号，以免影响空间的沉稳大气。低调华丽感，以及延续流畅是整个空间设计的主题，设计师打造出不失豪华气度的居所，且兼顾古典与时髦。本案的风格魅力在于摒弃了古典的繁琐，吸纳古典的华丽，以欧式惯用的元素点缀：不论格调相同的家具、帘幔、地毯、外罩等装饰织物，还是颇有欣赏和收藏价值的传统餐具、茶具等器具，都给古典风格的家居环境增添了端庄、典雅的贵族气息，并呈现出古典欧式在中国的全新生命力。

GRACEFUL AND COMFORTABLE

优雅舒适

The leading actor on the ground is no doubt the carpet. Colors of the carpet are bright or dim here and there in a so great way well matching with the surrounding environment. It integrates space and vision forging a graceful and comfortable atmosphere.

地面的主角不折不扣是属于地毯的,地毯颜色或深或浅,因地制宜,与四周环境配合,使功能空间和视觉感受都呈现出整体统一性,营造出典雅与舒适的氛围。

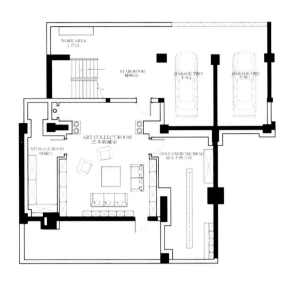

负二层

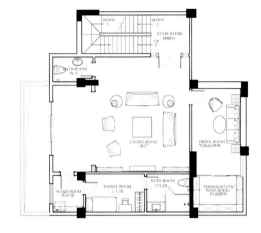

负一层

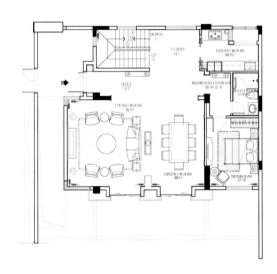

一层

二层

三层

SAKURA VILLA
樱花墅

TRADITIONAL
传统

FOLK
民风

GRACEFUL
雅致

RETURN TO CHINESE CLASSICAL
回归中式古典

Zhongliang Luofu Mountain & Water
Life Ecology Town, Huizhou
惠州罗浮山中量山水人间生态城

7
SAKURA VILLA
樱花墅

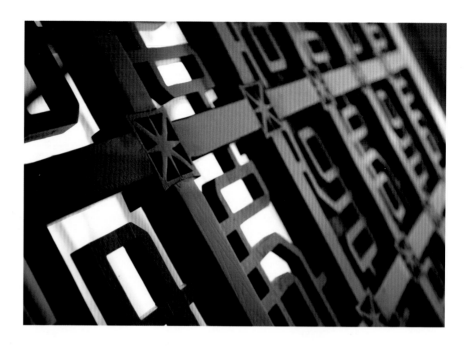

RETURN TO CHINESE CLASSICAL
回归中式古典

Project name Zhongliang Luofu Mountain & Water Life Ecology Town, Huizhou	**项目名称** 惠州罗浮山中量山水人间生态城
Design theme Chinese Style	**设计主题** 中式
Building area 560 sqm	**建筑面积** 560 ㎡
Flat Type 5 bedrooms, 3 living rooms, 1 kitchen and 7 restrooms	**户型** 五房三厅一厨七卫
Developer Shenzhen Zhongliang Investment Co., Ltd.	**开发商** 深圳市中量投资有限公司

The Zhongliang Luofu Mountain & Water Life Ecology Town combines together the Chinese elements and the modern workmanship. Through researching and using the local construction skills, the designer simplifies and abstracts the traditional decorations lines, analyzes and adopts modern materials, and integrates the grass, flower, bird and other folk subjects into the design, allowing it to be ancient yet modern, common yet graceful. The gray colored marbles are used as the indoor decoration materials, creating a special sense of beauty of space. The designer is good at using the complicated carving-patterned furniture, which are seldom used nowadays, to create a special space and set a great example, so that in the space, the Chinese elements can be found anywhere, which again shows the design principle of the designer of "combing past with present". The lamps are used to create shadows, and the unique light works perfectly with the indoor design and shows a great effect. The design of the lamps is inspired by Chinese traditional lamps, with traditional color and look, and by integrating them into modern element, the supervising effect is achieved.

会聚中国元素与现代工艺的罗浮山中量山水人间生态城，设计师通过对当地建构手法的研究与继承，传统装饰纹样的抽象简化以及现代材料的解析与运用，将卷草、花鸟等极具民风意向的主题融入室内设计，糅古释今，化凡为雅，使用了灰调大理石作为内部装修素材，营造出特殊时空美感。设计师善于运用如今少用的雕花繁复的家具塑造空间成为永恒典范，在空间内部随处可见的中式元素，再度验证了设计师刻意交织古今的设计巧思。用灯光打造出一系列的投影，独特的灯光效果与室内设计配合得天衣无缝，灯具设计灵感来源于中国传统灯具，传统的颜色与外形，融入现代感，达到意想不到的效果。

CHINESE ELEMENT
中国元素

COMBING PAST WITH PRESENT
糅古释今

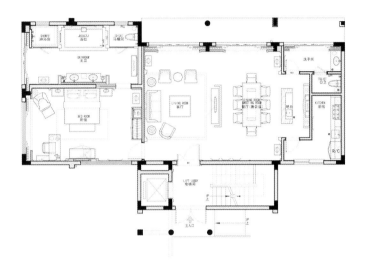

一层

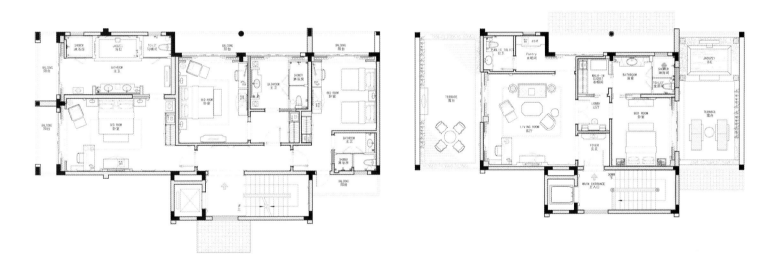

二层   三层

SAKURA VILLA
樱花墅

CONCEPT
抽象

CLEAR
澄澈

ATMOSPHERE
格调

A TASTEFUL LIFE
有品位的生活

A2, Building 2, Oriental Vision Phase I
锦绣御园一期 2 号楼 A2

A TASTEFUL LIFE
有品位的生活

A2, Building 2, Oriental Vision Phase I	锦绣御园一期 2 号楼 A2
Neoclassic	新古典
174 sqm	174 ㎡
4 bedrooms, 2 living rooms, 1 kitchen and 2 restrooms	四房二厅一厨两卫
Shenzhen Jinxiu Jiangnan Investment Co., Ltd.	深圳市锦绣江南投资有限公司

Hegel has defined aesthetics as "the emotional show of the concept". The exhibits in the space, no matter in what pattern, are relative to design. The designer uses the neoclassic lines to plan the relation between design and aesthetics. In a space with a certain base color, making use of the order of the lines, integrating the sense of layer of champagne golden, through the decoration of dots, diamonds, squares and curves, and basing on the luxurious design, the vitality of the space has been enhanced and the visual focus of the space has been enriched. The lamp and curtain decoration bring the ancient atmosphere, and the diamond-shaped glass wall, metal-texture artworks, feature sofa and classical furniture also give people the feeling of being in a warped space. The designer aims to present a clear and tasteful style in the space, and creating an appreciable atmosphere out of life.

黑格尔给美下了一个定义：美的理念就是感性的显现。空间内的陈设品，不管以什么样的形式出现，都与设计息息相关，设计师借由新古典的线条，规划设计与美的关系。在有着既定基调的空间里，利用线性规律的秩序，融入香槟金的层次，透过点状、菱形、方格、曲线的铺陈，依循华丽的设计演绎，增强空间的活力与丰富空间的视觉焦点。灯饰与窗帘带来古典的氛围，菱形玻璃墙面、金属质感艺术品、真皮沙发和古典家具从形式上给观者以时空错位的感觉，设计师旨在表达清晰而有格调的室内空间品位，酝酿生活之外的可鉴赏性氛围。

NEOCLASSIC LINES
古典风格

The designer uses the neoclassic lines to plan the relation between design and aesthetics.

设计师借由新古典的线条，规划设计与美的关系。

灯饰与窗帘带来古典的氛围，菱形玻璃墙面、金属质感艺术品、真皮沙发和古典家具从形式上给观者以时空错位的感觉。

The lamp and curtain decoration bring the ancient atmosphere, and the diamond-shaped glass wall, metal-texture artworks, feature sofa and classical furniture also give people the feeling of being in a warped space.

灯饰与窗帘带来古典的氛围，菱形玻璃墙面、金属质感艺术品、真皮沙发和古典家具从形式上给观者以时空错位的感觉。

APPRECIABLE ATMOSPHERE
赏析格调

SAKURA VILLA
櫻花墅

CULTURE
底蘊

LUXURIOUS
奢华

GRACEFUL
雅致

GET IMMERSED IN THE AGE
沉醉于岁月的沉淀

A1, Building 3, Oriental Vision Phase I
锦绣御园一期 3 号楼 A1

GET IMMERSED IN THE AGE

沉醉于岁月的沉淀

Project name	项目名称
A1, Building 3, Oriental Vision Phase I	锦绣御园一期 3 号楼 A1
Design theme	设计主题
New Chinese style	新中式
Building area	建筑面积
175sqm	175 ㎡
Flat type	户型
4 bedrooms, 2 living rooms, 1 kitchen and 2 restrooms.	四房二厅一厨二卫
Developer	开发商
Shenzhen Jinxiu Jiangnan Investment Co., Ltd.	深圳市锦绣江南投资有限公司

The feature of the Chinese style is simple and symmetrical, with rich culture and graceful style, and it is considered as the symbol of having high aesthetic value and social status. The Chinese flavor visual impression is moved into the design, so the beauty of east goes into the space not sharply, but tenderly, in a soft and tender manner, and through combination of the past and present, the imagination is inspired. The Chinese style creates the déjà vu in the space, just as the square shaped lines at the gate giving people a feeling of entering a noble family. The gate has a red color, generous and steady, just like the long history of China. The designer combines the green plants together with the red gate, making the space changeful and interesting, as if people are emerging in the changes of the ages. The bedrooms and living rooms are well separated from each other; the ancient Chinese style furniture is placed in the space, looking noble and elegant, yet luxurious and simple; the natural marble tiles show the generous life attitude of the host.

中式风格的特点是简朴而对称，富有丰富的文化性，且格调高雅，是具有较高审美情趣和社会地位的象征。中国风情的视觉印象转移至设计中，东方美并非张扬地侵入空间，而是以一种洗练轻巧的姿态存在其中，透过历史性画面与现实场景的重组，触动内心联想。中式风格的造型语言，似曾相识地重现于空间里。至玄关处开始的回形纹，以方正的形态唤起在大宅院游走的印象。大宅门沉稳的红色调不经意地流露，一如中国亘古悠久的历史，设计师特意将富有生命力的绿色植入其中，顺应空间的抑扬变化，变幻出仿佛是岁月沉淀般的丰富层次。动静分区的人性化设计，将生活的片段完美切割；古香古韵的中式家具，陈列于广阔空间，典雅高贵，不失奢华却也显得低调内敛；天然的大理石切面瓷砖，彰显主人大气磅礴的生活态度。

SO THE BEAUTY OF EAST GOES INTO THE SPACE NOT SHARPLY, BUT TENDERLY, IN A SOFT AND TENDER MANNER, AND THROUGH COMBINATION OF THE PAST AND PRESENT, THE IMAGINATION IS INSPIRED

以不露痕迹的方式将传统东方的唯美融入空间设计,并通过温柔灵感与融汇古今的想象力完美呈现

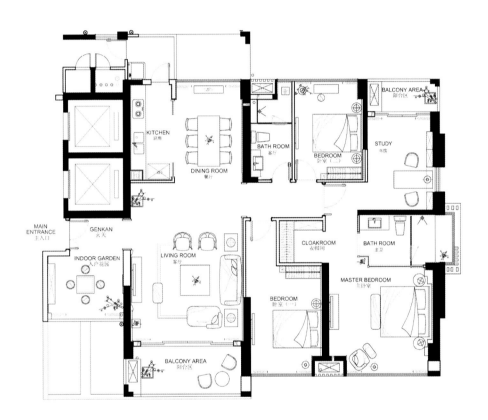

SAKURA VILLA
樱花墅

SPECIAL
独特

PRACTICABILITY
实用性

VISION
视觉感

ART OF LAYERED SPACE
层次空间艺术

Xiamen Yongjian – Dingshang Capital Villa
厦门永建·顶尚商墅

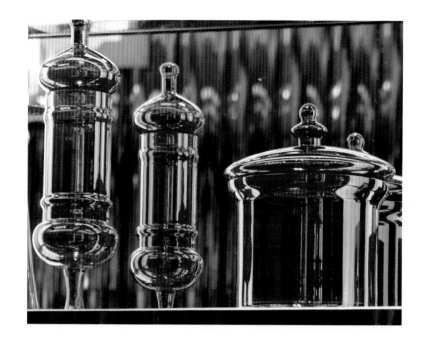

ART OF LAYERED SPACE
层次空间艺术

Xiamen Yongjian – Dingshang Capital Villa	厦门永建·顶尚商墅
Hong Kong style	港式
215 sqm	215 ㎡
2 bedrooms, 2 living rooms, 1 kitchen, 3 restrooms, 1 shop and 1 office	两房两厅一厨三卫一商铺一办公室
Xiamen Yongjian Real Estate Investment Co., Ltd.	厦门永建房地产投资有限公司

It's a new housing style with both the functions of residence and commerce. Special arrangements are adopted in special flat types, where the practicability and sense of vision are mainly taken into consideration, so that at the same time when fully satisfying the function of the space, the visual requirements can also be satisfied. The space is designed with the concept of "open and assembly", thus the functions of residence and commerce can be changed freely. The stairs are surrounded by glass, leading the way of the space, where the light can pass through into the space. The light reflective materials used in the room allow the villa to be highly unified visually. Under the light, the strong sense of space can be created.

新型的住宅形式，兼具商业和住宅功能。对于特殊户型采取特殊的安排，主要考虑实用性与视觉感，充分满足空间功能性的同时，完善视觉的要求。空间设计本着"开放、集成"的理念，因此生活与商业用途可自由更改。楼梯由玻璃包裹，引领空间走向，光线尽情倾斜，无所不在。室内反光材质的运用，使整个别墅空间达到视觉上的高度统一。在光线的穿梭下，营造出强烈的空间感。

DAZZLING

The first floor has a height of 3.6 meters, with shining shop windows, which may be the flagship shop of the crystal family, fashion or photo studio, as you like.

3.6米挑高首层,橱窗闪闪发亮,是水晶世家的旗舰店,也可以是时装、影楼等连锁平台。

With desks, office chairs and manager's room in it, it is like a high-end office building, creating a bridge for the art, advertisement or network enterprises.

电脑桌、办公椅、经理室……犹如高档写字楼,为艺术、广告、网络等企业架起宏伟蓝图。

DELICATE
AND SOFT

The third floor is the kitchen, dinning room, and the fourth floor is the main bedroom, simple yet luxurious, like the reception hall of the city, or the mansion of the leader.

三层厨房、餐厅，四层主卧套房，简约演绎奢华，是城市会客厅，也是领袖的都市行馆。

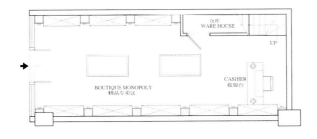

一层

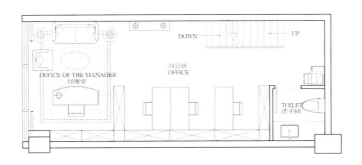

二层

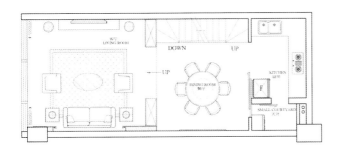

三层

四层

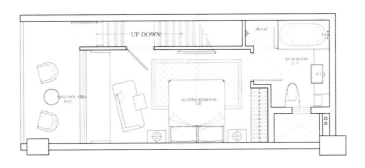

五层

SAKURA VILLA
樱花墅

SIMPLIFY
简洁

CHARM
魅力

ATMOSPHERE
格调

CHARM SOURCES FROM CULTURE
文化感传承风韵

Unit 2-A, Building 2, Shum Yip Yuyuan
深业御园 2 号楼单位 2-A

11
SAKURA VILLA
樱花墅

CHARM SOURCES FROM CULTURE
文化感传承风韵

Project name Unit 2-A, Building 2, Shum Yip Yuyuan	项目名称 深业御园 2 号楼单位 2–A
Design theme Modern Chinese style	设计主题 现代中式
Building area 89 sqm	建筑面积 89 ㎡
Flat Type 4 bedrooms, 2 living rooms, 1 kitchen and 2 restrooms	户型 四房二厅一厨二卫
Developer Shum Yip Southern Land (Holdings) Co., Ltd.	开发商 深业南方地产（集团）有限公司

The ancient Chinese style gives a brand new explanation to the Chinese ancient elements with a fashionable and modern design method, creating an artistic space combining oriental culture and international taste, thus the different spaces can show the different artistic feelings in a unified style. The design is full of elements of ancient architectures in southern China, together with the uncommon furniture with Chinese red, the Chinese atmosphere has been enhanced. To the contract, the wooden color in the room is full of sense of culture, and with the Chinese style lines throughout the space, the Chinese atmosphere has been further emphasized. Meanwhile, the hangers with the cracked patterns enhance the sense of space and enrich its layers. However, the Chinese style based on modern life has to be lightened to simplify the details while showing its charms, and thus the new elements and materials are adopted.

中式复古设计风格，将中国古典元素以时尚、现代的设计手法全新演绎，营造出兼具东方文化与国际品位的艺术空间，不同的室内空间在统一的中式风格中呈现出不同的艺术感。设计充满了南方古典建筑的设计元素，带有不同于常的中国红装饰，具有浓郁的中国氛围。与之相对应的木色强调了室内空间的文化气质，贯穿空间的中式纹样，加重了空间的中式氛围，冰裂纹的挂件加大空间感与丰富空间层次。但以现代生活为基调的中式，其本身必然有着相宜的淡化处理，体现风韵的同时，各式细节层面有着很大程度的精简，并增添了新元素和材质的运用。

CHINESE ATMOSPHERE
中国风

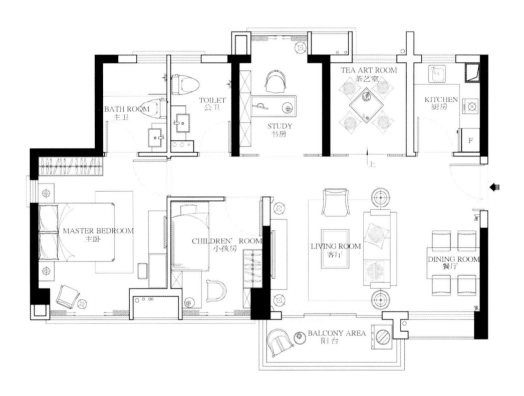

SAKURA VILLA
樱花墅

EXCELLENCE
优质

FASHIONABLE
潮流

SPECIAL
独特

FUN ABOUT BALANCE
有关平衡的乐趣

Kunming Vanke – Baisha Runyuan Type C
昆明万科·白沙润园 C 户型

FUN ABOUT BALANCE
有关平衡的乐趣

Project name Kunming Vanke – Baisha Runyuan Type C	项目名称 昆明万科·白沙润园 C 户型
Design theme Hong Kong style	设计主题 港式
Building area 170 sqm	建筑面积 170 ㎡
Flat Type 4 bedrooms, 2 living rooms, 1 kitchen and 3 restrooms	户型 四房二厅一厨三卫
Developer Kunming Vanke Real Estate Co., Ltd.	开发商 昆明万科房地产开发有限公司

12 SAKURA VILLA
樱花墅

Hong Kong is a very special place for people in China and even the whole world. In the past centuries, the eastern and western culture melted together in Hong Kong, creating different classics and excellence. As for furniture culture, Hong Kong also has its unique style – Hong Kong style, which has great value in the furniture design industry. Hong Kong style furniture is mainly into modern style, most of which have cold color, metal texture and lines, creating a simple but fashionable atmosphere. The whole space in this design is in modern Hong Kong style, with fashionable feature adopted as the material; on the wall of the dinning room, it is the night scene of the metropolis, and the color melts into the space at a far away place, making the whole space very "Hong Kong". If we say architects are stones who can sing, the decoration is his voice, smart and dynamic, which people and light dancing in it. Here we can see the game between shadow and exhibits, where the light passes through the top of the exhibits, colorful and changeful, just like a hold style Hong Kong made film.

香港在中国，乃至世界范围内，都是一个极为特殊的地域，过去的数百年间，中西文化在此剧烈碰撞，交融汇聚，衍生了各式各样的经典与精彩，而对于家居文化来说，香港也有着其独特的风格，"港式"一词在家居设计行业举重若轻。"港式"家居设计潮流多以现代为主，大多色彩冷静，以金属色和线条感营造简洁而不失时尚的气氛。整个空间是现代港式风格，材质上使用了时尚的皮质物料，餐厅墙上的大幅都市夜景图，色调在远远的地方与空间融合，使整个空间时尚而港味十足，倘若说建筑是会唱歌的石头，那软装就是他的声音，灵性舞动，人与光线都在其中自然舞动。这里有光影与陈设的游戏，光线轻巧地从陈设品的边界掠过，几许多变，几许阴影，其中韵味如一部老式的港产电影。

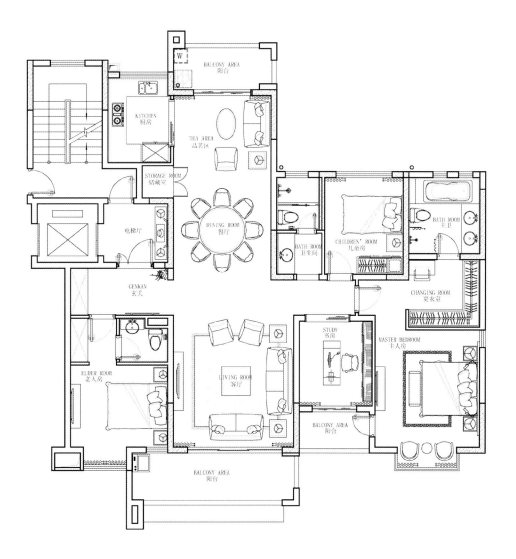

SAKURA VILLA
樱花墅

ROMANTICISM
浪漫唯美

BALANCE
均衡和谐

ANCIENT
古典雅致

TELL AN EXPECTED STORY

叙说一个向往的故事
Changcheng–Longwan
长城·珑湾

13 SAKURA VILLA
樱花墅

TELL AN EXPECTED STORY
叙说一个向往的故事

Project name Changcheng – Longwan	项目名称 长城 • 珑湾
Design theme Old Shanghai	设计主题 老上海
Building area 295 sqm	建筑面积 295 ㎡
Flat Type 4 bedrooms, 5 living rooms, 1 kitchen and 2 restrooms	户型 四房五厅一厨二卫
Developer Shanghai Shenchangcheng Real Estate Co., Ltd.	开发商 上海深长城地产有限公司

The old Shanghai always represents a style of romanticism. Western culture can usually be found in the design and life at that time. The essence of this design lies in charms, balance, melody, ancient and solemn, and at the same time in the vision and affection. The old Shanghai style can be defined as a collection of different styles. Whether the wallpaper with bird and flower patterns or the decorations with different flavors, they are all the results of the combination of different styles, bearing the unique charms of those years, and attracting the eyes of different people. Just like the glass color painting at the entrance of the stairs, as the expression of the space, its color and history can always draw people's attentions. The designer integrates the details, materials and colors together, and pays a special attention to details, while creating the materials full of ancient feelings, and removing the boundary between what contains in the article and the article itself. The details are supporting the entire space, and the superposition of details makes a unified article.

老上海是一种浪漫主义的风格。这一时期的设计与生活，都是一种东西方文化的糅合，整个设计的精华就是风情、平衡、韵律、怀旧和庄重，同时又兼具视觉与情感的高度，这一种老上海风格，可定义为一种风格的拼贴。棕褐色的墙裙、花鸟壁纸、风情万种的饰品便是将各种风格融合的结果，有着那个特定年代的独特风韵，牢牢抓住了观赏者的眼球，就像楼梯口彩绘的玻璃画，作为一个空间最丰富的表情，它的色彩、它的过去让人忍不住会去关注。设计师将细节、材料和颜色整合成一个整体，对于细节尤为关注，精心制作了复古感浓烈的物料，模糊了建筑物本身的和其包含物的界限，这些细节支撑起整个空间，而细节的叠加形成一个统一的实体。

ROMANTICISM
STYLE

浪漫主义格调

It is hard to find the ear that the space is in at the first glance. This is because that mood and connotation are embodied in details as long lasting flowing streamlet. Only as time goes by and carefully can you find the story.

并非一眼就能看出空间所处的年代，因为它只在细节中传达情绪与内涵，缓若细水长流般，用心用时间，才能发现那些故事。

EXPECTED STORY

期待传奇

Existence of stairs seems to give more stories. What it leads to is not limited to other space.

楼梯的存在，仿佛就是为了增添故事性，它所通往的地方，并非仅限别处的空间。

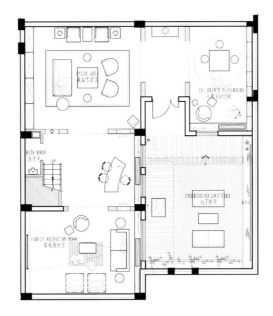

负一层

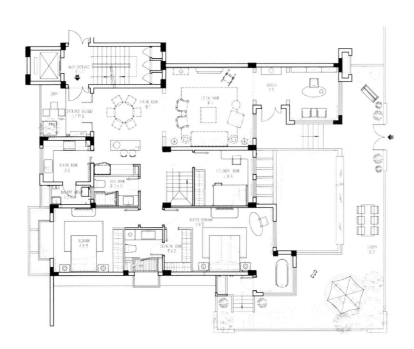

一层

SAKURA VILLA
樱花墅

SIMPLENESS
简洁

IMPRESS
铭刻

TRANSPARENT
澄清

SIMPLE BLACK & WHITE
简约黑白协奏

A1, Building 1, Oriental Vision Phase I
锦绣御园一期 1 号楼 A1

SIMPLE BLACK & WHITE
简约黑白协奏

Project name	项目名
A1, Building 1, Oriental Vision Phase I	锦绣御园一期 1 号楼 A1
Theme of Design	设计风格
Simple European Style	简欧
Building area	建筑面积
161 sqm	161 ㎡
Unit Type	户型
3 bedrooms, 2 living rooms, 1 kitchen and 2 restrooms	三房二厅一厨二卫
Developer	开发商
Shenzhen Jinxiu Jiangnan Investment Co., Ltd.	深圳市锦绣江南投资有限公司

The designer puts its appreciations to life in the design. He values simpleness and gives up complication, and looks for the inspiration in the purest colors. The whole design shows a relation witch strong contract between black and white while maintains the perfect balance, combining all the exhibits together. The black and white squares highlight the space and create the European style simple lines in the space, while the contract between black and white shows a modern feeling which differs from the European style. The glass door and gate make the entire space transparent, with clear layers while extending the eye sight. The wall adopts marble material, witch light gray color and unique lines, showing the stableness of the black and white and bringing dynamic feeling to the space, to make people impressed.

设计师投入一种对生活的鉴赏态度，崇尚简练与平衡，摒弃繁复，在最纯粹的色彩中寻找灵感，赋予空间黑白的表情。整个设计拥有一种黑白强对比但又维持着微妙的平衡关系，将所有的陈设品融入其中，黑白格的踊跃是空间活跃点，创建出的空间有着欧式的简单线条，黑白的对比，有一种区别于纯欧式的现代感。玻璃门和大门使得整个室内空间显得十分通透，延展了视线的同时，层次分明。墙面选择大理石材，偏灰的色调搭配独特的纹理沉淀空间格调，牵引了黑白的内敛沉稳，使得视觉灵动富有微妙的动感，给人以感性的触动。

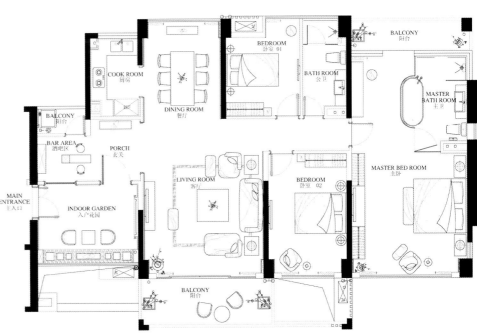

SAKURA VILLA
樱花墅

PEACE
宁静

CHANGEFUL
丰富

EXPERIENCE
意会

SIMPLE ELEGANCE OF BLUE
蓝色简约之雅

Unit 2-B, Building 2, Shum Yip Yuyuan
深业御园 2 号楼单位 2-B

SIMPLE ELEGANCE OF BLUE
蓝色简约之雅

Project name Unit 2-B, Building 2, Shum Yip Yuyuan	项目名称 深业御园 2 号楼单位 2-B
Theme of Design Modern European style	设计主题 现代欧式
Building area 89 sqm	面积 89 ㎡
Flat Type 4 bedrooms, 2 living rooms, 1 kitchen and 2 restrooms	户型 四房二厅一厨二卫
Developer Shum Yip Southern Land (Holdings) Co., Ltd.	开发商 深业南方地产（集团）有限公司

15 SAKURA VILLA
樱花墅

The white color makes the light and wall in the space elegant and graceful. When there is sunlight, the space is full of conflicts between brightness and darkness, with clear layers, which represent the ability of the designer in controlling place, light and direction. The classic blue will never fade with time, and no matter how the world changes, people always seek for relaxing their body and mind and the feeling of peace. The blue color is always the theme of a home, and is also a test for the designer. The material used in the room is the key in the design, while the slight changes in lines and the lights can bring people rich visual and sensual experience. By adopting light reflective and absorbing materials, smooth and rough materials, the soft and hard textures and natural and artificial textures can be presented. Different materials used in the room make the blue room rich and changeful.

白色的存在使得整个空间的光线与墙面显得尤为雅致，阳光流动，空间变得充满了明暗的对比，层次分明。高度考验了设计师对地域、光线、方位的高度掌控力。蓝色回归，经典的色彩终究不会被潮流淹没，无论外界如何变幻，人们最渴望的还是让身心彻底放松，找回宁静的感觉，而蓝色从来不辱使命。蓝色调注定是家居永恒的主题，蓝色的搭配非常考验设计师的水平，室内材质、形体的对比是关键，微妙的纹理变化、光影的变幻给人带来丰富的感官效果，采用反光与吸光材质、光滑与粗糙材质、柔软与坚硬质感、天然与人工触感的对比彰显出艺术效果。迥异的材质，让蓝色居室的表情更富有变化。

Scattered pearls, corals and shells are the gifts from sea. The sense of sea is full of the space featuring pure and natural.

散落的珍珠、珊瑚、贝壳……是大海馈赠的礼物，海的气息弥散在空间之中，纯粹而悠然。

ELEGANT AND GRACEFUL

While facing the sea, spring scenery is expressed in various ways. What you see here is no doubt one the most outstanding ones.

面朝大海,春暖花开的意境,有着多种不同的表达形式,而这里呈现的,无疑是最为杰出的形式之一。

典雅大方

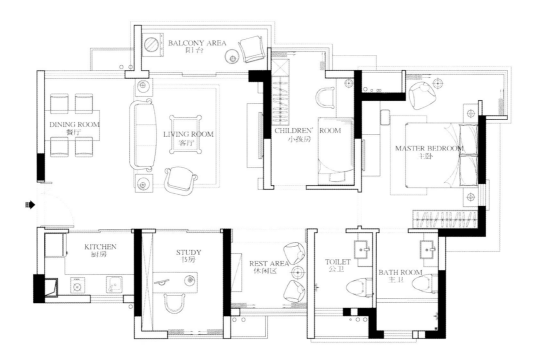

SAKURA VILLA
樱花墅

UNIQUE
别致

SIMPLE
简约

COMBINATION
融汇

FROM DREAM TO REALITY
梦想照进现实

Shangdongwan Type B-A

上东湾 B-A 户型

FROM DREAM TO REALITY
梦想照进现实

Project name	项目名称
Shangdongwan Type B-A	上东湾 B-A 户型
Design theme	设计主题
Modern style	现代
Building area	建筑面积
88.32 sqm	88.32 ㎡
Flat Type	户型
4 bedrooms, 2 living rooms, 1 kitchen and 2 restrooms	四房二厅一厨二卫
Developer	开发商
Shenzhen Peng Guang Da Industry Co., Ltd.	深圳市鹏广达实业有限公司

The designer carefully shows the unique space, where the exhibits are well collocated with each other, avoiding a dull visual effect. The element of design is simple and unified, and the color and material are harmonious with each other. By making good use of the perfect combination of the points, lines and planes, and with the sense of power of the lines and the pure color, the designer has created a powerful visual sense in different functional spaces. The use of the light color enlarges the space, generous but yet with rich details. Fashion is a necessary element in modern style, full of the warmness and funs of the modern life. You can say that inspiration is very important in this design, it is not a creation on purpose, but a flash of intuition, then the perfect combination of dream and reality is created.

设计师细心推敲出空间的别致新颖，空间陈设的相互渗透避免了单调的视觉效果，设计元素简单统一，色调与材质搭配和谐。设计师利用点、线、面的完美组合，利用线条的力量感、高纯度的色彩，在简单明了的各功能空间营造出强有力的视觉感。浅色的大面积使用扩大了空间，大气不失细节。时尚元素是现代风格呈现的必要介质，充满了现代生活的温馨感，且趣味十足。可以说，灵感在这里尤显重要，它无关刻意的生造，只求灵光一闪，便实现了梦想与现实的完美链接。

CRYSTAL CLEAR
晶瑩剔透

Existence of mirror extends your visual space. Light blue color grants mild elegance to the space, not to please but in a connotive way to show a gentle and satiated posture.

镜面的存在延伸了视觉空间,而淡淡的蓝色给予了空间以淡雅,不取悦却也不含蓄,呈现柔和的饱满姿态。

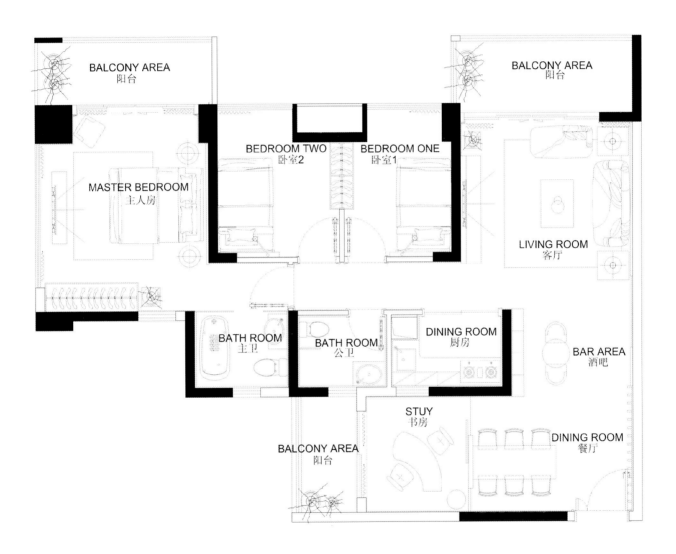

SAKURA VILLA
樱花墅

BALANCE
平衡

FRESHNESS
新鲜

MYSTERIOUS
神秘

A SPACE WITH SANSKRIT
梵音弥散的空间

Kunming Vanke – Baisha Runyuan Type D
昆明万科·白沙润园 D 户型

A SPACE WITH SANSKRIT
梵音弥散的空间

17 SAKURA VILLA
樱花墅

Project name	项目名称
Kunming Vanke – Baisha Runyuan Type D	昆明万科·白沙润园 D 户型
Design theme	设计主题
New Chinese style	新中式
Building area	建筑面积
133 sqm	133 ㎡
Flat Type	户型
3 bedrooms, 2 living rooms, 1 kitchen and 2 restrooms	三房二厅一厨二卫
Developer	开发商
Kunming Vanke Real Estate Co., Ltd.	昆明万科房地产开发有限公司

The new Chinese style deign is a combination between Chinese and modern style, creating a balance between the traditional and modern aesthetics. It boasts both the new and old culture, containing Chinese style elements and at the same time the top-end fashion. The designer uses dark and light colors to bring more layers to the space and make it vivid and brilliant. The wooden floor makes the space not so modern. The black wooden frames are adopted throughout the space, in doors, pictures, cabinets, windows, etc. The well allocated lines and the correspondence of the space are all representations of the space with Chinese flavor, allowing modern people to experience traditional Chinese culture. The highlight is in the dinning room, where a group of green plants is placed, showing the feeling of peace and freshness, and together with the Buddhism painting on the wall, they bring the breath full of elegance, coziness, nature and Sanskrit, showing the mysterious oriental culture.

新中式风格的作品，是当下中式与现代交融的混搭风格，创造传统与现代美学之间的平衡。它会聚新旧文化的宽阔气度，既蕴涵中式元素，又是走在潮流最尖端的时尚风格创作。设计师运用深浅不同的颜色为空间注入更多层次，空间灵动生辉。整齐的木质地板冲淡了现代感。黑色木质边框的使用贯穿所有空间，门框、画框、柜台、窗框，线条的契合性，空间的转折与呼应都是一种对中国情境空间的表达，符合现代人对于传统中式文化的向往。特别点是在餐厅之中，一组绿色植物的放置，清新迎面，衬出一片宁静，与墙面的佛像壁画映衬出一种含蓄雅致，充满沉静与超然的禅意气息，展示东方文化的神秘性。

SUBTLE AND ELEGANT
含蓄雅致

QUIET AND ALOOF
沉静超然

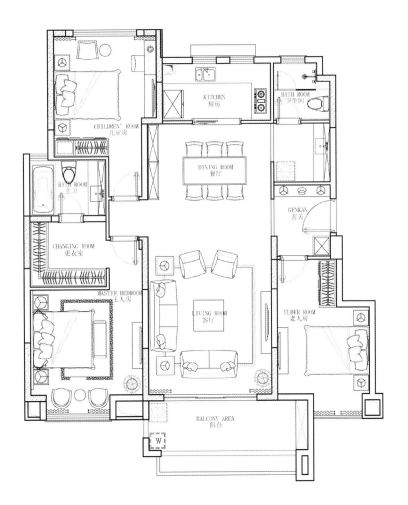

SAKURA VILLA
樱花墅

DYNAMIC
活跃

MEANINGFUL
意蕴

FUNCTION
功能

THE ART OF LINE
线形的艺术

Kunming Vanke – Sales Office of Baisha Runyuan
昆明万科·白沙润园售楼处

THE ART OF LINE
线形的艺术

Kunming Vanke – Sales Office of Baisha Runyuan	昆明万科·白沙润园售楼处
Modern Chinese style	现代中式
1016 sqm	1016 ㎡
Reception hall, Exhibition area, Negotiation area, Waiting area, VIP room, Office area	接待门厅、展示区、洽谈区、签约等候区、VIP室、办公区
Kunming Vanke Real Estate Co., Ltd.	昆明万科房地产开发有限公司

In this design the lines play the most important role in the space. The warped and dynamic lines make the space more meaningful together with the soft color and texture of the solid wood. The extension of lines and the design of indirect lighting become the boundary between the bedrooms and living rooms, and at the same time show the sense of art that is out of reality. The folding screen in the space makes several private rooms and smaller spaces, which perfect the function and value of the space.

全案以线条为空间表情，略显错落的动态线条搭配实木的温润颜色及触感，增加空间的主题意义，线性的延续，加上间接照明的设计，有效地成为空间中动、静态的中介，顺势勾勒出超脱于现实的艺术设计感。阻断屏风的存在，兼顾私密性与空间感，在整体空间中滋生众多小空间，完善其功能价值。

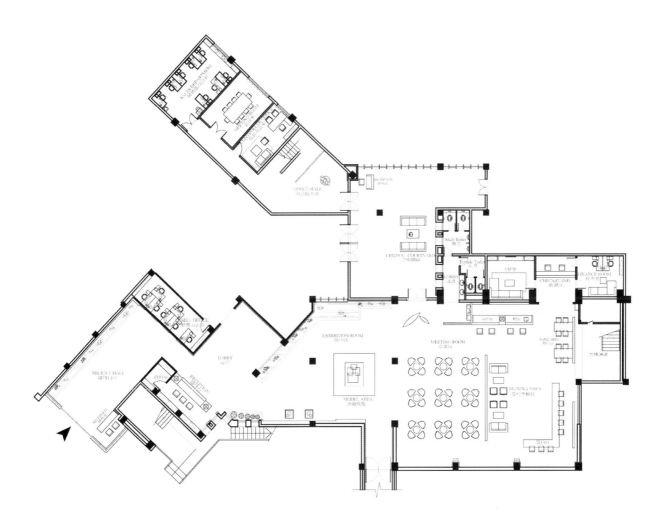

SAKURA VILLA
樱花墅

WARM
温馨
EXPERIENCE
体验
EXPRESSION
描绘

THE MAGIC OF MOSAIC
马赛克幻觉魔法

1-C, Unit B, Building 1, Shum Yip Yuyuan
深业御园 1 号楼 B 单位 1-C

19
SAKURA VILLA
樱花墅

THE MAGIC OF MOSAIC
马赛克幻觉魔法

Project name
1-C, Unit B, Building 1, Shum Yip Yuyuan

Design theme
Modern

Building area
37 sqm

Flat Type
1 bedroom, 1 kitchen and 1 restroom

Developer
Shum Yip Southern Land (Holdings) Co., Ltd.

项目名称
深业御园 1 号楼 B 单位 1-C

设计主题
现代

建筑面积
37 ㎡

户型
一房一厨一卫

开发商
深业南方地产（集团）有限公司

When you step into the room, you will feel as if you are in a space of magic. The visual feeling in the space gives people a very unique experience. The black and white mosaics in the space represent a message that is not ignorable. The wallpaper has the texture similar with leather, which creates a warm and fashionable effect in the entire space. The expression of modern comes from the understanding to the time, and the conflicts between the black mirror surface and the wallpaper extend the visual filed and at the same time unify the visual feeling and represent the romance of the host. In other spaces, such as the bedroom, kitchen or restroom, the same concept is adopted, where the black and white mosaics' ability in extending the space is emphasized.

走进室内，就像穿越进了魔幻空间，空间的视觉感给人以十分独特的体验，黑白马赛克在空间里的存在，传达了无法忽视的组合排列讯息。墙纸有着类似于皮革的质感，这种材质在整个空间营造出一种温暖而时尚的效果。现代感的肆意表达来源于对时代的深刻了解，黑色的镜面与墙纸之间的强对比，扩张了视觉领域的同时，在视觉感受上又极其统一，抒发了空间主人的浪漫情怀。再往其他空间去，卧室、厨房、盥洗室设计都遵从相同的原理，一贯而下强调黑白马赛克的空间扩张力。

SPACE OF MAGIC
魔幻空间

The visual feeling in the space gives people a very unique experience. The black and white mosaics in the space represent a message that is not ignorable.

空间的视觉感给人以十分独特的体验，黑白马赛克在空间里的存在，传达了无法忽视的组合排列讯息。

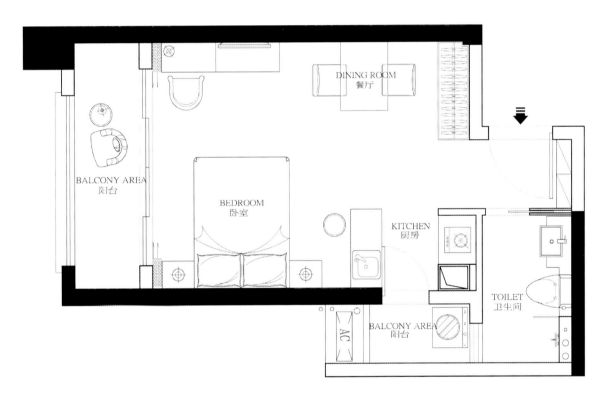

SAKURA VILLA
樱花墅

INSPIRATION
灵感

MEDITERRANEAN
地中海

VISUAL
视觉

THE BLUE OF OCEAN
海的蓝在此飞扬

Kunming Vanke – Baisha Runyuan Type E1
昆明万科·白沙润园 E1

20 SAKURA VILLA
樱花墅

THE BLUE OF OCEAN
海的蓝在此飞扬

Project name
Kunming Vanke – Baisha Runyuan Type E1

Design theme
Mediterranean

Building area
92 sqm

Flat Type
3 bedrooms, 2 living rooms, 1 kitchen and 2 restrooms

Developer
Kunming Vanke Real Estate Co., Ltd.

项目名称
昆明万科•白沙润园 E1

设计主题
地中海

建筑面积
92 ㎡

户型
三房二厅一厨二卫

开发商
昆明万科房地产开发有限公司

How to give the space the style of ocean, is what the designer wants to find out. In Kunming, the designer has found the inspiration. He directly places the colors to visually unify the ocean indoor and the blue sky outdoor, thus the feeling space is enhanced. In a series of spaces with Greece white color, the blue colors of different depths are used, and the selection of material and colors give people the feeling of being bright, fresh and clean. The indoor space adopts the wind-eroded decorations and flannel cloths, controlling the melody of color of the space and creating the atmosphere of the Mediterranean. The expression of blue makes the entire space light and relaxing feeling, and the fresh painting on the wall extends the visual boundary of the space. The small decorations are spread in different places of the room, like the gifts from the ocean, giving people a feeling as if they are collecting shells on the beach.

如何赋予空间以海洋的风格，是设计师最为关注的一点，幸好在昆明的地域氛围之下，设计师找到了灵感，以最为直接的色彩铺排，使室内的海与室外的蓝色天空形成视觉上的一体化，空间感被加强，一系列希腊式的白色空间里汇集了各种层次的蓝色，材料、颜色和质地的选择给人一种明亮、新鲜和纯净的感觉。内部空间采用风蚀过的饰品和粗绒的布料，掌控空间的色彩节奏，营造地中海的氛围。蓝色的率性表达，感官上给予整个空间以轻快，墙面的清新壁画纵深感突出，延展空间的视觉界限。小饰品的存在，犹如大海的馈赠散落在房间的各个角落，给人以沙滩之上拾贝的雀跃。

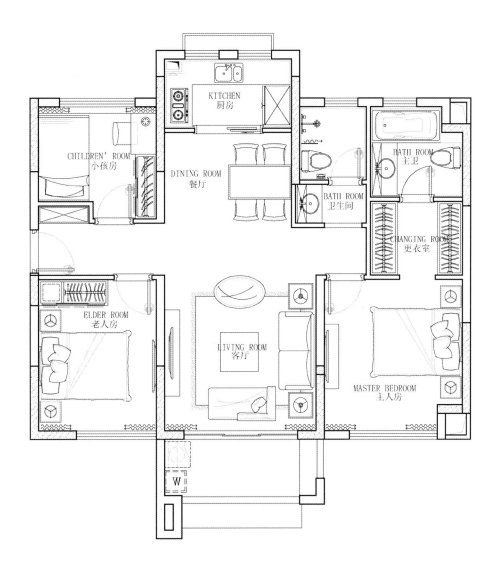

SAKURA VILLA
樱花墅

HARMONIOUS
和谐

ELEMENT
元素

LUXURY
奢华

DRAMATIC BEAUTY WITH
HIGH CONTRACT
强对比的戏剧性美感

1–B, Unit B, Building 1, Shum Yip Yuyuan
深业御园 1 号楼 B 单位 1–B

DRAMATIC BEAUTY WITH HIGH CONTRACT

强对比的戏剧性美感

Project name	项目名称
1-B, Unit B, Building 1, Shum Yip Yuyuan	深业御园 1 号楼 B 单位 1-B
Design theme	设计主题
Modern	现代
Building area	建筑面积
64 sqm	64 ㎡
Flat Type	户型
3 bedrooms, 2 living rooms, 1 kitchen and 1 restroom	三房二厅一厨一卫
Developer	开发商
Shum Yip Southern Land (Holdings) Co., Ltd.	深业南方地产（集团）有限公司

The dramatism can be emphasized by the strong conflicts between the high reflective materials and warm cloths, and the green, as the main color of the space, is throughout the entire space, making it fresh and bright, and easing the conflicts. The large area of green is limited in the light cherry color of the floor, looking warm and implied. The material of the furniture adopts cream colored feature, which is harmonious with the environment. Hangers with curve lines are decorated on the green decorative wall, and the matte painting gives them the noble feature. The theme pattern of the space is the Eiffel Tower picture on the wall in the main bedroom, with customized luxury and brilliance. The entire style is unified with different materials and decorations, and the harmony in design concepts between the background and the overall space has been the highlight of this design, where the diversified elements show that the sport and art can coexist in harmony.

利用高反光材质与温暖布料的强对比来凸显戏剧性，而作为空间主题色彩的绿色贯穿整个空间，让整体风格清新明快，适当缓和冲突。大面积的绿色被控制在地板浅色的樱桃木色中，温暖而含蓄。家具面料采用米色的皮质材料，与整体环境相协调，绿色的装饰墙面点缀着卷纹的挂件，亚光的亮漆使其具有尊贵的特色。空间的主题图案是主卧的艾菲尔铁塔背景墙，定制的奢华，呈现耀眼的光芒，整体设计风格在不同的材质和不同的装饰语言上得到了统一，背景设计方式的考虑与设计的构思协调统一成为设计的亮点，多重元素结合体现运动与艺术的和谐共处。

21
SAKURA VILLA
樱花墅

SAKURA VILLA
樱花墅

VITALITY
活力

INSPIRATION
灵感

ATMOSPHERE
格调

MAKE EVERYTHING AT EASE
让一切回归安然

B2, Building 1, Oriental Vision Phase I
锦绣御园一期 1 号楼 B2

22 SAKURA VILLA
樱花墅

MAKE EVERYTHING AT EASE
让一切回归安然

Project name	项目名称
B2, Building 1, Oriental Vision Phase I	锦绣御园一期 1 号楼 B2
Design theme	设计主题
Modern	现代
Building area	建筑面积
94 sqm	94 ㎡
Flat Type	户型
2 bedrooms, 2 living rooms, 1 kitchen and 1 restroom	二房二厅一厨一卫
Developer	开发商
Shenzhen Jinxiu Jiangnan Investment Co., Ltd.	深圳市锦绣江南投资有限公司

Green is a color full of vitality. In this design, green is adopted as the main color; in the clean and bright space, the fresh and fashionable green becomes the highlight, representing the designer's inspiration of seeking and searching coziness. From the living room to the dinning room, and then to the bedroom, different romantic scenes have been created. Meanwhile, the clean lines becomes another highlight in this design; whether the TV wall, the ceiling or the surface of the wardrobe, are all extending this kind of atmosphere. Based on sensibility, the reasonability is also adopted to maintain the magnificent structure of the space, while the dramatic conflicts between deep and shallow bring connotation to the design. By experiencing this sensible atmosphere, it is easy to feel at ease.

绿色是充满生机的色彩，本案大胆采用绿色为主色调，在干净明快的空间布局下，以或清新、或时尚的绿色创意亮点，呈现出设计师追寻与探索的惬意设计灵感，从客厅、餐厅到卧室，勾勒一个个浪漫的场景。同时，干净利落的线条也是本案的一大亮点，无论是电视墙、天花还是衣橱外的剖面，都在延续这种理性基调。在感性的柔和之下，以理性维持空间的大气格局，浅层与深层的戏剧冲突为内涵的孕育提供了可能。汲取这种可以感知的意境，体会安然自然水到渠成。

SEEKING AND SEARCHING

寻寻觅觅

In the clean and bright space, the fresh and fashionable green becomes the highlight, representing the designer's inspiration of seeking and searching coziness.

在干净明快的空间布局下，以或清新或时尚的绿色创意亮点，呈现出设计师追寻与探索的惬意设计灵感。

SAKURA VILLA
樱花墅

CHARACTERISTIC
独特

VIVID
生动

MAGNIFICENT
旖旎

LIFE IN FALSE OR TRUE
虚实之间理解生活

1–D, Unit B, Building 1, Shum Yip Yuyuan
深业御园 1 号楼 B 单位 1–D

LIFE IN FALSE OR TRUE
虚实之间理解生活

23 SAKURA VILLA
樱花墅

Project name 1-D, Unit B, Building 1, Shum Yip Yuyuan	**项目名称** 深业御园 1 号楼 B 单位 1-D
Design theme Modern Simple European style	**设计主题** 现代简欧
Building area 76 sqm	**建筑面积** 76 ㎡
Flat Type 3 bedrooms, 2 living rooms, 1 kitchen and 1 restroom	**户型** 三房二厅一厨一卫
Developer Shum Yip Southern Land (Holdings) Co., Ltd.	**开发商** 深业南方地产（集团）有限公司

The unique characteristic of this design lies in the lines that separate the walls. While emphasizing the independence of the space, it never cuts off people's eye sight. It is a sublimation of the modern European style, yet has a different understanding to the traditional one. The design shows the diversified functions and flexibility of the furniture, maybe warm, or tender, or romantic, casting beautiful reflections with the color decorations in the room; with the harmonious and natural arrangement, the spaces are well related with each other, and the eye sights can move freely in different spaces. The space has a magnificent structure, with clean and clear design; the warm lines of the wallpaper bring the lines of the space closer to each other, giving people an atmosphere of being home, noble and luxurious, comfortable and cozy. The designer focuses on the overall space so that the different functional spaces can show their unique feature in unity, which make the home life more vivid.

纹路隔断墙面的出现是本案的独具匠心，强调空间独立性的同时，并不阻断视觉的延伸，一种现代欧式的升华，却不同于以往对简欧的理解。整个设计充分演绎出了家居功能的跳跃与灵便性，或温馨，或柔和，或浪漫，与室内精彩纷呈的装饰和摆设相映成趣，搭配自然和谐，空间有机相连，视线在空间中自由流畅。空间结构开阔大气，设计手法干脆利落，墙纸的温暖纹路缓和空间线条的疏离感，烘托出家的氛围，使其奢华高贵气质立显，而又不失舒适感。设计师善于把握全局，使不同的功能分区在统一中呈现出自己独特的个性，使家居氛围更为活跃。

UNIQUE CHARACTERISTIC
独具特色

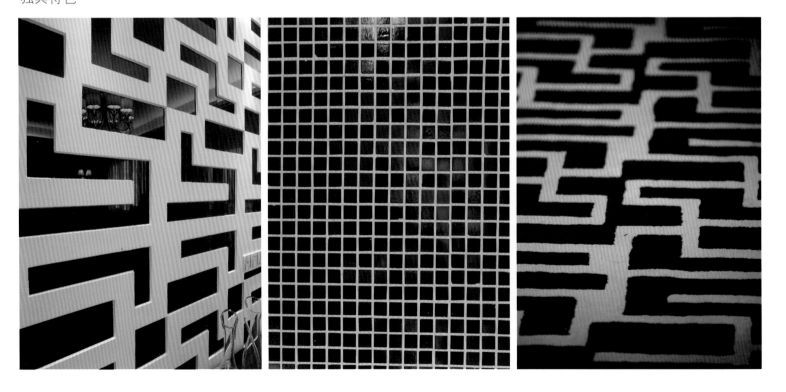

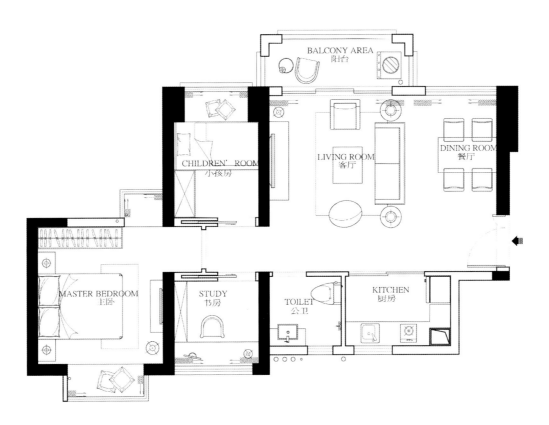

SAKURA VILLA
樱花墅

EYE-CATCHING
璀璨

COMFORTABLE
优裕

MODERN
现代

FUN OF TENDERNESS
柔和的趣味

Shangdongwan Type B-F
上东湾 B-F 户型

FUN OF TENDERNESS
柔和的趣味

Project name Shangdongwan Type B-F	项目名称 上东湾 B-F 户型
Design theme Modern	设计主题 现代
Building area 48.08 sqm	建筑面积 48.08 ㎡
Flat Type 2 bedrooms, 2 living rooms, 1 kitchen and 1 restroom	户型 二房二厅一厨一卫
Developer Shum Yip Southern Land (Holdings) Co., Ltd.	开发商 深圳市鹏广达实业有限公司

The eye-catching cherry red and white can be a good complement to each other, and with the light purple in between, a comfortable, modern, convenient and humanized space is represented. This is an attempt of the designer, trying to find out the support to the theme in the colors of the same kind – and this is never a difficult thing. In each room, the light, color and shadow are taken into consideration, and the soft colors are brought into the relaxing and elegant funs, thus to satisfy the most rigorous demands.

醒目的樱花红与白色相互协调和弥补，搭配淡紫色作为中间色调，充分体现舒适、现代、便捷的人性化空间。这是设计师的一个高调尝试，试图在同色系色彩中寻找到主题的支撑，所幸这并非难事，每一个居室都体现了对光、色、影的综合考虑，柔和的色彩带入了轻松雅致的趣味，从而从根本上解决了某种近乎苛刻的需求。

24
SAKURA VILLA
樱花墅

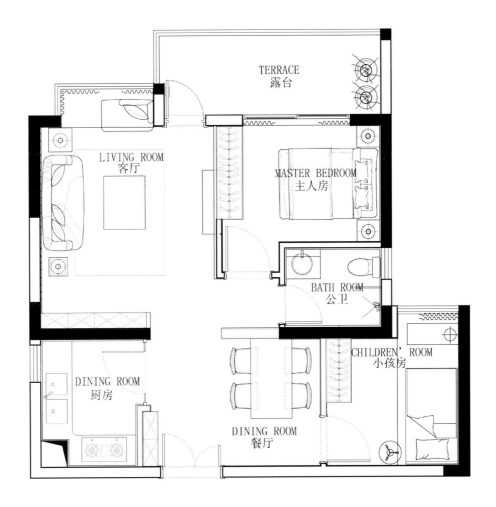

CONCEPT

T SPACE 概念空间

If you are touched by
The plain and the persistent Chinese red

If you have been indulged in
Historical atmosphere overflowing everywhere

It is the bearing of culture
It is also the inheritance of culture

The elapse and regression of time
Both are here.

是否被打动
那一抹质朴且坚毅的中国红

是否已沉迷
那于任何一处洋溢的历史气息

既是文化的承载
亦是文化的传承

时间的流逝与回归
皆在此处

Colorful life

缤纷生活

There is surprise anytime in life

Just as in cold winter

You can see caltx canthus in blossom

It is also seems

Having found in heart

The purest Chinese pattern

Red over the brick wall

Echoes with poem during childhood

Spreads texture of memory

人生时刻拥有惊喜

就像在最寒冷的冬季

遇见最傲艳的腊梅

也像在此

找寻到存在于内心深处

最纯正的中国格调

砖墙上探头的红

呼应儿时的诗篇

舒展了记忆的纹理

HISTORICAL ATMOSPHERE
复古格调

Taste a cup of classy red wine
The thing that makes people cheerful and indulged

Is never only the taste at the moment
But also the legendary psalm behind

Vineyard under the sunlight of the distant land
Is provided with inborn attracted specialty

Oak barrel deeply hidden in the cellar
Is telling stories with unique charm

品尝到一杯上等的红酒
令人愉悦沉醉的

绝不只是当下的滋味
更有那背后的传奇诗篇

遥远国度阳光下的葡萄园
天生具备吸引人的特质

深藏于地窖之中的橡木酒桶
讲述着独具魅力的故事

CHEERFUL AND INDULGED

愉悦与迷恋

UNIQUE CHARM
独特魅力

In quietly elegant and composed light
Jumping musical notes seem to be there

Freely sing in the space
Sprinkle emotions that can not be hidden

It is not tenderness or quietness
It is not passion or high spirit

The wonder does not lie in telling and listening
But in the inner resonance of person and space

淡雅安然的灯光里
仿若可以看见跃动的音符

于此空间畅意协奏
挥洒无法掩饰的情绪

并非柔和或静谧
也非激情或昂扬

美妙之处不在倾诉与聆听
在于人与空间的内心共鸣

TELLING AND LISTENING

娓娓道来 侧耳倾听

JUMPING MUSICAL
跳跃的音符

It is only the lingering charm of oriental wisdom
That can sufficiently touch the heart

All things met, seem, known and felt,
Are like bright-moon in autumn
Filled with clear and peaceful artistic conception

Joy in life is hard to come by
While encounter and harmonious stay are more difficult to seek

仅是东方智慧的余韵
就足以牵引内心

所遇所见所知所感
皆如秋之明月
充满清朗安宁的意境

人生喜悦难得
更难寻相遇而安

ONE THOUSAND YEARS PEACE FOR ONE THOUSAND YEARS WORDLESS PITY COUNTLESS STARING

一千年 一千年平和 静默无言 长久凝视

The beauty that green is shining through
Stretches vigorous vitality of nature

Charm of nimble dancing and
Delivers the joyous singing in the depth of the forest

Living in prosperous place for long
We forget another conversational ability

Only through communication between different hearts
We can experience miracle and mystery of everything

绿意盎然之美
舒展自然的蓬勃生机

轻灵舞动之魅
传递森林深处的欢鸣

久居繁华之地
忘却了另外一种对话能力

不同心灵之间的交流
方能体会万物的神奇与神秘